HEALING WITH COCONUT

pil
Publications International, Ltd.

Written by Jacqueline B. Marcus, MS, RDN, LDN, CNS, FADA, FAND

Photography from Shutterstock.com

Copyright © 2018 Publications International, Ltd. All rights reserved. This book may not be reproduced or quoted in whole or in part by any means whatsoever without written permission from:

Louis Weber, CEO
Publications International, Ltd.
8140 Lehigh Avenue
Morton Grove, IL 60053

Permission is never granted for commercial purposes.

ISBN: 978-1-64030-179-5

Manufactured in China.

8 7 6 5 4 3 2 1

TABLE OF CONTENTS

INTRODUCTION p. 04

COCONUT BASICS p. 08

HOW DO COCONUTS HELP HEALTH p. 13

I HEART COCONUTS p. 23

FAT TO FIGHT FAT p. 35

USE YOUR BRAIN p. 45

COCONUT THE BEAUTIFUL p. 55

Introduction

Most people wouldn't recognize the term *"Cocos nucifera."* But the fruit of the palm *Cocos nucifera*—the incredible coconut—is a multidimensional and tremendously useful member of the plant kingdom. It can help you to prevent and treat some common health problems, nourish your body and groom your hair and nails, supercharge your metabolism, and much more.

Coconuts provide concentrated energy in their "meat," refreshing hydration in their water, hardy fibers for household necessities like brushes and twine, and a hard shell that can be converted into charcoal or used for various handcraft applications—and all these uses come in one compact and economical package!

Coconut Origins

Coconut palms may have originated in Malaysia. This conclusion is based on the number of species of insects around the world that are associated with coconut palms, but botanists do not fully agree. A broader viewpoint is that coconuts originated somewhere in tropical Asia. Part of what makes their origin so difficult to pinpoint is that coconuts are so hardy. It is speculated that coconut seeds probably floated to different parts of the globe before humans transported them to different cultures.

Some early evidence of coconut use comes from the Marquesans, who have lived on a group of small islands located in the South Pacific a few degrees south of the equator since around AD 300. Coconuts were thought to have been consumed in a variety of ways throughout the Marquesas. Immature coconuts were cut open and their creamy interiors were used to nourish children. Coconut "cream" was extracted from freshly grated coconut meat, which was also combined into other dishes. Marquesan men grated coconuts with

an apparatus that looked like a sawhorse. The coconut blossom was also a favorite food source. Its sap, called *jekamai*, was heated to prevent fermentation and the sweet syrup of the blossom was used to flavor dishes.

Europeans first recognized the existence of coconuts in the Middle Ages, though Arab traders might have introduced coconuts to eastern Africa much earlier. The explorer Marco Polo (left), a Venetian merchant traveler and explorer who engaged in an epic round-trip Asian journey that lasted 24 years, is said to have discovered coconuts in Java and the Nicobar Islands in Southeast Asia.

While the coconut was reportedly mentioned in ancient eastern Indian documents, the Asian geography that is west of India is not truly suitable for its growth—nor is the south of China. Excursions by sailors and tradesmen both north and south brought coconuts into new lands.

During pre-Columbian times in Central America, the earliest Spanish invaders are said to have found coconuts growing along the western coast of Panama. They subsequently introduced it to Puerto Rico.

The Portuguese explorer Vasco da Gama (right), who was the first European to reach India by sea and link Europe and Asia by an ocean route, located coconuts on an island off Mozambique in the 15th century. Then Portuguese explorers introduced coconuts to Brazil in the 16th century. However, it wasn't wasn't until the 19th century that cultivation spread throughout Florida.

Today, the main areas growing and exporting coconuts are Brazil, the Caribbean, India, Indonesia, Malaysia, Mexico, Papua New Guinea, the Philippines, Sri Lanka, and East and West Africa.

Coconut Geography

Coconut palms thrive along seascapes. Coconut palms are usually tall and prefer light, salty, and sandy conditions with airy soil, achieved by the ebb and tide of the neighboring seawater. The tree has shallow, widely spread roots that allow the coconut palms to gently sway, while the huge leaves (often 20 feet long with massive midribs for stability) also provide needed shade from the penetrating tropical sun for the coconut palms.

Harvesting and Using Coconuts

The coconut is the world's largest—and some say the most important and useful—nut. It is not just humans who think so: When not harvested by humans, coconuts may be picked by local monkeys right off of the palm trees.

In modern coconut plantations where dwarf cultivars of coconuts are grown, coconuts may be harvested directly from the ground with hooked knives attached to bamboo poles.

The coconut requires a sharp strike of force to split it open for its meat and water. A spike or machete is traditionally used to strike the end of the coconut and remove the husk. The coconut husk contains useful fibers known as *coir* that are used for coconut matting, among other purposes.

The "meat" or *copra*, located within the interior of the coconut, is firm and creamy white in texture and snowy white in appearance.

Also inside of the coconut is a hollowed center that is filled with a sweet, watery liquid called coconut water. Coconut water is sometimes referred to as coconut "milk," but it is really a thin liquid. If the nut is "green" or young, once it is pried open, then the refreshing coconut water is released. A sharp instrument is generally pierced into the coconut "eyes" to enable the coconut water to be poured out or for a straw to be inserted. Once mature, the dried coconut meat (copra) may yield some additional coconut water.

In addition to coconut "meat" and coconut water, coconuts produce a wide array of products that include coconut "cream," coconut "milk," and coconut "oil." Coconuts provide nourishment both as a food and as a beverage, uel for cooking, and a vessel for serving. Coconuts also produce basket materials, chemicals, medicines, textile fibers, thatching, timber, and other useful and valued products.

A Nutritional Nut

Aside from their most notable taste and aroma, coconuts have been valued for centuries for their nutrients—particularly for their fats with energy and health-enhancing benefits. The calories contained within coconuts from fats, carbohydrates, and proteins are impressive.

What is particularly impressive about the nutrients in coconuts are the types of fats. Within 133 grams of total fat, one medium coconut contains about 118 grams of saturated fats (590 percent of the Daily Value), 1.5 grams of polyunsaturated fats, and 6 grams of monounsaturated fats. While these amounts may seem high, there are good explanations about these fats and why they are considered to be so important for diet and health. A thorough explanation of dietary fats and cardiovascular disease and weight control are found in later sections of this book.

While it is rare that a person consumes an entire medium coconut at one time, even in smaller amounts coconuts provide a proportional array of these nutrients.

One medium coconut (397 grams) contains about 1,405 calories, with:

- 133 grams of total fat (204 percent of the Daily Value [DV] based on a 2,000 daily calorie diet)
- Zero cholesterol
- 60 grams of carbohydrates (144 percent of the Daily Value)
- 36 grams of dietary fiber
- 25 grams of naturally occurring sugars
- 13 grams of protein (26 percent of the Daily Value)

Other nutrients in one medium coconut include:

- Calcium (5 percent of the Daily Value)
- Vitamin C (21 percent of the Daily Value)
- Iron (53 percent of the Daily Value)
- Vitamin B-6 (10 percent of the Daily Value)
- Magnesium (31 percent of the Daily Value)

Coconut Basics

What is basic about the coconut? Hardly anything at all! Coconuts are very simple, yet very complex in character. Understanding the inner workings of coconuts is key to appreciating their impressive characteristics and benefits.

Unique Classification

Coconuts are uniquely classified as a seed, fruit, and nut. This is because a coconut is the stone or seed of a *drupe*, or fleshy fruit (the seed-bearing structure of a flowering plant), that is capable of reproducing. The word "drupe" is derived from the Latin word *drūpa* or *druppa*, which translates to overripe olive.

The fruit grows on the *Cocos nucifera*, treelike palms that are part of the Arecaceae palm family and the only species of the Cocos genus. These palms are more closely related to grasses than they are to nut trees. However, the coconut is also considered a nut, which is a fruit that is composed of a hard, tough shell that encapsulates an edible kernel.

Structure

Coconuts are encircled by a smooth, deep tan hard outer covering that encloses the husk and a thick and fibrous layer of fruit and milk. The outside husk of a coconut is called the *exocarp*, or outer layer. The exocarp is smooth and very strong. It is green to reddish brown in color, and grays as a coconut matures. In some varieties of coconuts, the exocarp is ivory in color. The husk has three indented "eyes" at one end of the coconut.

The husk is lined with a thin brown skin called the *testa* that firmly adheres to the hollow kernel that is filled with coconut water. The *mesocarp* (fleshy middle layer) is thick and filled with coarse brown fibers. Within the hairy husk is the *endocarp* (a hard, woody layer that surrounds the seed) that is filled with coconut meat and milk.

Development

Coconuts originate and develop over the course of a year. A coconut is filled with liquid after about four months and attains its full size around five months when its meat becomes jellylike.

Immature coconuts are about five to seven months old and are filled with sweet coconut water that contains about two percent sugars. Immature coconuts also contain fragile, gelatinous, moist meat that is composed of sugars and other carbohydrates and water.

Coconuts are considered to be mature after about 11 to 12 months of growth. They may weigh about 2 to 5 pounds and contain about 15 percent water, with one-quarter of their total weight composed of creamy, gelatinous, textured coconut meat. As coconuts continue to age, the meat becomes more solid and lightly fibrous.

As the coconut meat develops its fatty, firm, and white characteristics, the coconut liquid becomes less plentiful than within immature coconuts. At this stage of development, mature coconut meat is comprised of about 45 percent water, 35 percent fat, 10 percent carbohydrates, and 5 percent protein. The liquid provides a sweet and refreshing drink—often called coconut juice or simply coconut water.

Dietary Staple

Coconuts have been a staple food and beverage in the diets of many cultures for thousands of years, in Polynesia, Hawaii, and Sri Lanka.

Polynesians

The Pukapuka and Tokelau of Polynesia, who live on atolls near the equator, consume diets that are high in saturated fat, but low in dietary cholesterol and sucrose. Coconut is their main source of calories. Vascular disease is uncommon among these Polynesians and there is little evidence of their high saturated fat intake having harmful effects.

There are two distinctively different forms of coconuts that are consumed in Polynesia: the *niu kafa* and *niu vai*, Samoan names for these Polynesian varieties.

The niu kafa coconuts of Polynesia (as the coconuts that made their way to Florida) are triangular and oblong with large fibrous husks. This is in comparison to the niu vai coconuts that are rounded and brightly colored green, red, or yellow when unripe, with plenty of sweet coconut water inside.

Hawaiians

The first Hawaiian Island settlers who arrived from the Marquesas Islands around AD 300 to 400 are thought to have been of Polynesian descent. They likely brought Niu, the coconut that was elevated from just an ordinary food to a sacred tree in the Hawaiian Islands.

Niu is depicted in mythical art and verbal lore as a magical tree that is an image of *Ku*, the Hawaiian ancestor and the link to their heritage and life itself. Niu was valued cargo since it was considered to be the most useful plant of the tropics. There are stories of Hawaiian Islanders who survived months of drought by consuming coconut water. Ancient Hawaiians were said to be strong and sturdy from their native diet, which was considered one of the best in the world for the time.

Sri Lankans

The Veddas, or "Forest People," are indigenous people of Sri Lanka. The majority of Sri Lankans are of the Sinhalese race that migrated from India around two thousand years ago.

Coconut palms were very important to the Veddas for their steady source of food and drink, as well as materials that were used for building fires, eating utensils, rope, and tools.

The majority of dietary fat in the traditional Vedda diet was derived from coconuts and wild game that are both high in saturated fats. In the 1980s the Veddas were studied to determine how their high-fat diet affected their health. Their rate of cardiovascular disease was very low. Heart disease and stroke were also virtually nonexistent in Kitava, Papua New Guinea, where coconuts, fish, fruit, and tubers are the main dietary staples.

While these indigenous populations survived and thrived on coconut meat and coconut water, it is important to note that their lifestyles were very different from typical North American lifestyles today, with much more activity and a less-processed diet without many of the stresses of modern society.

While it is easy to make correlations between coconut consumption and decreased or absent diseases of affluence (such as cardiovascular disease, diabetes, or hypertension), one needs to keep the isolation and location of population groups such as the Veddas in mind. Still, the low cardiovascular disease rate despite the high saturated fat consumption of the Veddas is an impressive trait worth studying.

Types and Advantages of Coconut Oil

Coconut oil is generally liquid at room temperature, though this characteristic may change depending upon its country of origin. However, it solidifies at temperatures below that, so refrigerated or frozen coconut oil will look quite dense.

After processing, coconut oil is almost tasteless, so it is commonly used in baked goods, confections, cooking oils, and margarines for its great versatility. There are several types of coconut oil, mostly distinguished by the types or lack of processing.

Unrefined coconut oil, also called "pure" or "virgin" coconut oil, is made from fresh coconut meat instead of dried. The coconut oil is extracted by either a quick-dry method or through a wet-milling process.

Quick drying dries the coconut meat rapidly; then the coconut oil is pressed out mechanically. *Wet milling* first processes the coconut meat into coconut milk. Then the coconut milk is separated from the coconut oil by boiling, centrifuging, fermenting, straining, or by the use of enzymes.

Because both of these extraction methods are quick, the coconut oil does not require additives or bleaching. Both processes also retain more flavor since they do not expose the coconut oil to high temperatures.

Refined coconut oil is made from dried coconut meat, or copra. (Some copra-based refined coconut oils are referred to as "RBD oils" since they are **r**efined, **b**leached, and **d**eodorized.) The coconut meat is bleached (generally not by a chemical process, but by filtration) and treated to reduce any bacteria. However, the drying process itself may produce contaminants.

A high heat process is then used to extract the coconut flavor and aroma. Sometimes chemicals are used to extract as much oil as possible. Sodium hydroxide may be added to increase shelf life. Refined coconut oil is sometimes partially hydrogenated, but this process may produce trans fats, which are disallowed in United States manufacture by the U.S. Food and Drug Administration (USFDA).

Virgin coconut oil is derived from the expeller pressing of coconut oil from dried (desiccated) coconut. It is not distinguished by an industry standard, much like in the olive oil industry; however, it is considered to be unrefined.

Extra-virgin coconut oil, similar to virgin coconut oil, bears little distinction to extra-virgin olive oil or any oil with an "extra-virgin" designation. It implies that this type of coconut oil has the highest quality and retains the original chemical composition and nutritive values of coconuts. Unlike olive oil there is no industry standard definition.

How Do Coconuts Help Health?

In this section, we'll look at five notable features of the incredible coconut that point to its healing properties.

1. Saturated Fats

The word is that fat is bad for us, and saturated fat especially, so how are the saturated fats in coconut different than other saturated fats in foods and beverages? This section discusses how, contrary to what the diet industry has the public believing, fat is needed in our daily diets. This nutrient is essential for processes such as brain function, energy production, and immune protection. Still, it is vital to consume the right kinds of fats and in the right amounts.

The "medium-chain triglycerides" (MCTs) that are found in coconuts (and discussed throughout this book—particularly in the next three sections) are considered to be more beneficial than some other kinds of fats and for good reason.

For instance, medium-chain triglycerides are more easily digested and immediately converted to energy, whereas "long-chain triglycerides" (LCTs), which make up the majority of the fat in our diets, are carried by the bloodstream and are not necessarily burned for energy. This section will distinguish among these types of fats and to illustrate why the coconut is a cut above other foods of its kind.

Chemically speaking, saturated fats are fat molecules that are saturated with hydrogen molecules. Practically speaking, saturated fats are fats that are filled to capacity and are difficult for the body to disengage.

Saturated fats have been on the nutrition and health radar for years due to their relationship to cardiovascular disease. The reason why saturated fats are discouraged is that they tend to raise the level of cholesterol in the bloodstream, which in turn may increase a person's risk of cardiovascular disease and stroke.

Conversely, replacing foods and beverages that are higher in saturated fats with those that are lower in saturated fats may lower serum (blood) cholesterol, improve lipid profiles in the blood, and conceivably even reduce the risks of cardiovascular disease and stroke. The *lipid profile* is a panel of blood tests that provides a screening tool for blood lipids that include cholesterol and triglycerides.

Polyunsaturated fats are fat molecules that have openings in their structures without hydrogen molecules. They tend to be lighter fats and are thought to be health enhancing. Polyunsaturated fats are sometimes made into saturated fats through the process of *hydrogenation*. Hydrogenation adds more hydrogen to fat molecules to make them more stable.

Consider canola oil. Through hydrogenation, canola oil could become hydrogenated into canola oil-based margarine, transformed from tub to stick. In the process, artificial trans fats are formed, which have been deemed unsafe by the U.S. Department of Agriculture.

Currently the American Heart Association (AHA) offers recommendations that Americans should aim for a dietary pattern that contains five to six percent of calories from saturated fat. If a person consumes 2,000 calories daily (the figure that the U.S. Food and Drug Administration uses to determine the Daily Values [DVs] of different nutrients), then this means that no more than about 120 calories or 13 grams of saturated fats should be consumed daily.

Also according to the American Heart Association, a daily diet pattern should emphasize fruits, vegetables, whole grains, low-fat dairy products, poultry, fish, and nuts, and limit red meats and sugary foods and beverages. The AHA also suggests that foods with high monounsaturated and/or polyunsaturated fatty acids should replace foods that are higher in saturated fatty acids.

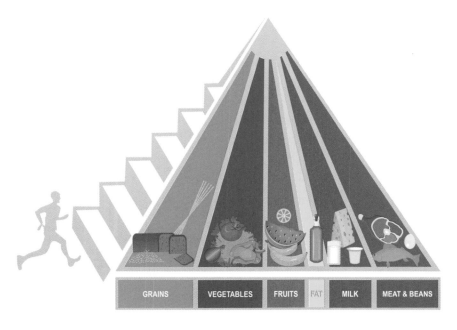

How Do Coconuts Help Health?

Some plant-based oils such as coconut oil, palm oil, and palm kernel oil do not contain cholesterol (a fatty substance that is also a risk factor in cardiovascular disease and stroke), but they do primarily contain saturated fats.

Coconuts, Saturated Fats, Health, and Heart Disease

What might be considered as an enigma to this discussion about the relationship between saturated fats and cardiovascular disease and stroke is the relationship of coconuts and cardiovascular and brain health that will be described more fully in later sections.

Out of the 133 grams of total fat in one medium coconut, 118 grams or roughly 88.7 percent is from saturated fats. This is approximately 590 percent of the Daily Value that is based on a 2,000 daily calorie diet. In comparison, the amount of polyunsaturated fat is only 1.5 grams and the amount of monounsaturated fat is only 6 grams.

Also in comparison, the amount of saturated fat in butter is about 64 percent of total calories and the amount of saturated fat in beef fat is about 40 percent of total calories, or about the same amount as in lard.

What is different about the saturated fats in coconut oil is that these fats tend to increase high-density lipoprotein cholesterol (or HDL—the "good" cholesterol). This is based on short-term studies that examined the affects of coconut oil on serum cholesterol levels. As of last year, the exact mechanisms by which coconut oil affects cardiovascular disease were not confirmed.

The saturated fat in coconut oil is mostly comprised of medium-chain triglycerides (MCTs). These medium-chain triglycerides are handled differently in the body than the longer-chain fatty acids that are found in dairy products, fatty meals, and liquid vegetable oils. This distinction is the next notable feature of coconuts.

2. Medium-Chain Triglycerides

Triglycerides are the most common type of fat in the body. They come from food; the human body also makes triglycerides. Once the triglycerides in foods and beverages are consumed and digested, they travel through in the blood and supply energy or are stored for future use. Foods and beverages that are higher in cholesterol and saturated fats tend to elevate serum (blood) triglyceride levels. High levels of blood triglycerides are frequently seen in people with heart problems, have diabetes, or are overweight.

There are different types of triglycerides that are differentiated according to the number and types of fatty acids, whether they are saturated or unsaturated, and their degree of unsaturation. Triglycerides are designed as short, medium, or long in length.

Short-chain triglycerides: Short-chain triglycerides are triglycerides that are generally about 2 to 5 carbon atoms in length. They are primarily absorbed through the portal vein during lipid digestion, much like medium-chain triglycerides are.

Medium-chain triglycerides (MCTs): Medium-chain triglycerides are triglycerides that are generally about 6 to 12 carbon atoms in length. The fatty acids in medium-chain triglycerides are generally broken off from their triglyceride structure without the need of bile (an alkaline fluid that aids in fat digestion) from the liver. Then the MCTs are "shuttled" via the portal artery directly to the liver without the use of chylomicrons (small, lipoprotein particles) for transport. This attribute makes medium-chain triglycerides desirable for their relatively quicker source of energy compared to long-chain triglycerides with less dependency upon the liver.

Long-chain triglycerides (LCTs): Long-chain triglycerides are triglycerides that are generally about 14 to 22 carbon atoms in length. Long-chain triglycerides usually require both bile and lipases (fat-digesting enzymes) to be emulsified and broken down into individual fatty acids in the small intestine. Then these fatty acids are reassembled and carried through the blood or lymph, used by the body as needed, or stored. The process is lengthier than medium-chain triglyceride digestion, absorption, and utilization and is dependent upon bile and lipases.

What makes the types of fat in coconuts different than the fats that are found in other foods and beverages is that practically all of the fat in coconuts are medium-chain triglycerides (MCTs), while the majority of fats that are consumed by humans are long-chain triglycerides (LCTs). This factor is responsible for the unique character and healthful properties of coconut oil and has many implications.

Medium-chain triglycerides are used in hospital formulas to feed very young infants, people who are critically ill, and those who may have digestive problems. This is because MCTs are easily digested, absorbed, and utilized to nourish the human body. Because of their quick conversion into energy, medium-chain triglycerides are also used by athletes, for weight management, and in the treatment of *epilepsy* in children. Epilepsy is a neurological disorder that is designated by recurrent episodes of sensory disturbances. When medium-chain triglycerides are used in conjunction with a prescribed ketogenic diet, symptoms appear to decrease. Later sections describe the ketogenic diet in more detail.

What makes the medium-chain triglycerides in coconuts different is that most of the fatty acids are saturated. Other than treating epilepsy, the critically ill, those with digestive disorders, and young infants, the medium-chain triglycerides in coconuts have shown effectiveness in cardiovascular risk prevention and health promotion. These attributes are opposite to the effects that other saturated fats may have on the human body and health.

This leads to another feature of the coconut that makes it different in good ways: the lauric acid content.

3. Lauric Acid

Lauric acid, a component of triglycerides, is saturated fatty acid with a 12-carbon chain, so it is considered to be a medium-chain fatty acid. Lauric acid comprises about one-half of the fatty acids in coconut milk and coconut oil (49 percent of total fat). It is also found in cow's milk (2.9 percent of total fat), goat's milk (3.1 percent of total fat), and human breast milk (6.2 percent of total fat).

Lauric acid increases total serum cholesterol due to its favorable increase in high-density lipoprotein cholesterol (HDL, or "good" cholesterol) that correlates with decreased risks of *atherosclerosis* (hardening of the arteries) and cardiovascular disease.

Both lauric acid and monolaurin also have demonstrated antimicrobial activity against some bacteria, fungi, and viruses. This is why, along with vascular disease, lauric acid may be used to treat a host of other conditions that include the avian flu, cold sores, the common cold, fever blisters, genital herpes and warts, the swine flu, and viral infections including influenza. And this is also why coconut oil, with its high percentage of lauric acid, is valued by some for its antibacterial, antifungal, and antiviral properties.

4. Coconut's Nutritional Value

Some foods and beverages boast that they are Mother Nature's perfect food. While no food or beverage is 100 percent perfect, the incredible coconut provides its share of nutrients in one convenient package.

First consider coconut meat, the edible white interior of coconuts.

- It is a storehouse of energy providing 283 calories per one cup of raw, shredded coconut meat.
- It provides carbohydrates with 12 tasty grams of carbohydrates per one cup.
- It has 7 grams of dietary fiber per one cup.
- It contains 27 grams of total fat with 24 grams of saturated fat and no cholesterol per one cup of raw, shredded coconut meat.
- It provides 3 grams of protein per one cup.
- It is low in sodium with only 16 milligrams of sodium per one cup.

Next consider coconut milk.

- It is a storehouse of energy with 552 calories per one cup of raw coconut milk.
- It provides carbohydrates with 13 grams of carbohydrates per one cup.
- It is a good source of dietary fiber with 5 grams of dietary fiber per one cup.
- It contains 57 grams of total fat with 51 grams of saturated fat and no cholesterol.
- It provides 5 grams of protein per one cup.
- It is low in sodium with only 36 milligrams of sodium per one cup.

Then consider coconut water.

- It is low in calories with only 46 calories per one cup of coconut water.
- It provides carbohydrates with 9 grams of carbohydrates per one cup of coconut water.
- It provides dietary fiber with 3 grams of dietary fiber per one cup.
- It contains zero grams of total fat, saturated fat, or cholesterol.
- It provides protein with 2 grams of protein per one cup.
- Note that it is higher in sodium at 25 milligrams per one cup.

This equates to plenty of energy from the calories in fats, carbohydrates, and proteins, plus dietary fiber and relatively little sodium with plenty of essential and supportive vitamins and minerals (to be discovered throughout this book). What other foods or beverages can boast all of these nutrients in these levels that are encapsulated into just one incredible coconut?

5. Body Benefits

What other fruit, nut, or seed has so many beauty properties that can be used in such a variety of different hygiene and beauty applications? Coconut oil can be used as a body oil, body lotion, body scrub, breath freshener, cuticle oil, makeup remover and makeup brush cleanser, dandruff treatment, deep conditioner, deodorant, frizz tamer, itch reliever, leave-in conditioner, lip balm, massage oil, night cream, under eye cream, shaving cream, stretch mark cream, and teeth whitener, among many other usages.

> A mixture of apple cider vinegar and coconut oil when combed through hair may be effective in ridding and preventing head lice. Just ask any parent of school-aged children who have been infested with head lice: they may agree that delousing the hair with a natural product such as coconut oil may be the alternative to harsh chemical treatments.

In comparison, avocadoes are technically classified as fruit since they are large berries that contain single seeds. Like coconuts, avocadoes have a host of beauty benefits. Avocadoes contain antioxidants carotenoids that include alpha- and beta-carotene, cryptoxanthin, zeaxanthin, and lutein. These antioxidants help to protect the skin from environmental damage from oxygen-free radicals, which may contribute to visible signs of aging, such as fine lines or wrinkles.

Plus, avocadoes contain vitamin C—also for healthy skin. Vitamin C is needed to create elastin and collagen that helps to bind the skin and maintain the firmness and structure of skin cells.

Another antioxidant that is found in avocadoes is vitamin E, which also helps to prevent free radical damage and may be protective against excessive sun exposure.

One of the biggest beauty benefits of avocadoes is their composition of monounsaturated fatty acids from their high oleic acid content. Monounsaturated fatty acids help to moisturize the epidermal layer of the skin to keep it hydrated and soft in appearance and are said to help decrease inflammation.

However, a virgin coconut oil-enriched diet was also shown to have a beneficial role in improving antioxidant status and preventing lipid and protein oxidation, or breakdown. Plus, the saturated fat content of coconuts has shining benefits.

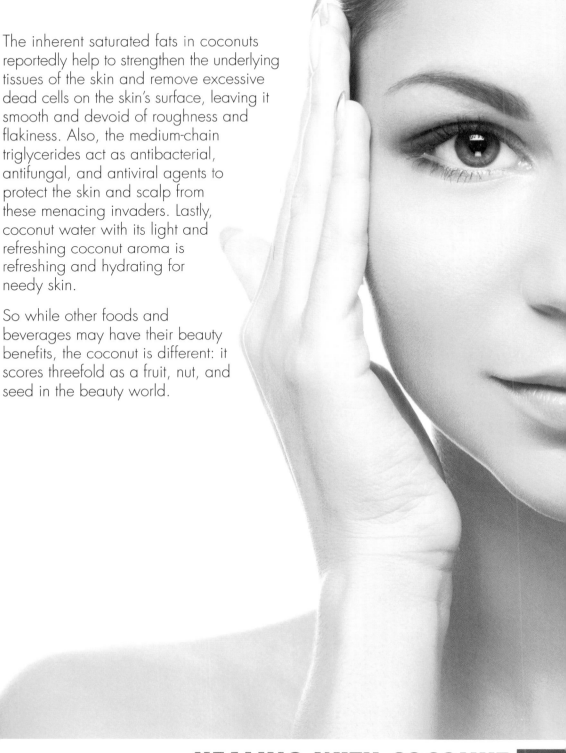

The inherent saturated fats in coconuts reportedly help to strengthen the underlying tissues of the skin and remove excessive dead cells on the skin's surface, leaving it smooth and devoid of roughness and flakiness. Also, the medium-chain triglycerides act as antibacterial, antifungal, and antiviral agents to protect the skin and scalp from these menacing invaders. Lastly, coconut water with its light and refreshing coconut aroma is refreshing and hydrating for needy skin.

So while other foods and beverages may have their beauty benefits, the coconut is different: it scores threefold as a fruit, nut, and seed in the beauty world.

Not Just for Health

The benefits of coconut don't end with your health. Let's take a quick look at some of the other uses and benefits of the amazing coconut.

- One surprising use of coconuts and what makes them a standout in the plant kingdom is their use to both whiten and shine the body and the home. Who would imagine that a natural substance such as coconut oil could whiten the stains on tooth enamel *and* shine leather shoes? Or that coconut oil may moisten and shine lips *and* polish wooden floors and furniture to a warm glow? Or that coconut oil may lighten simple bruises and scrapes *and* brighten cast-iron pans?

- It is difficult to walk into a bakery or perfumery without sensing the pleasantly sweet and tropical scent of coconuts. Likewise, the taste of coconuts is so distinct that it is hard not to identify the taste of coconuts in savory or sweet foods or beverages. This can be advantageous and disadvantageous. The scent and taste of coconut are so distinct that some may either have a love or dislike for these sensations. Some subtler smelling fruits, nuts, and seeds are too difficult to detect so assertively as coconuts.

- Coconut has many culinary applications in cooking and baking. The coconut meat, milk, and water are exceptionally versatile in appetizers, breakfasts, desserts, main dishes, side dishes, smoothies, soups, and stews. Coconut meat, milk, and water add aroma, fat, moisture, taste, and texture in common and surprising ways. For example, when shredded coconut is used in baked goods the flavor is full-bodied, moist, and tropical. And when toasted coconut tops fruit or curries, browning brings out the fatty, nutty coconut taste.

I Heart Coconuts

This section focuses on all the ways that the coconut has been shown to protect the heart. Contrary to past thinking, research has shown that the fat in coconuts does not contribute to heart disease and on the contrary, may protect the heart and prevent heart disease.

The coconut is thought to increase levels of high-density lipoproteins (HDL, or "good" cholesterol) and lower levels of low-density lipoprotein (LDL, or "bad" cholesterol) and may also help to lower blood pressure.

In this section the roles that the coconut plays in reaping heart-healthy benefits are discussed.

Cardiovascular Disease

For more than 50 years cardiovascular disease has been one of the major causes of death in the United States. The underlying cause of cardiovascular disease is atherosclerosis, which is a condition that is characterized by plaque, or fatty deposits that are lodged within the arterial walls. Plaque interferes with blood flow to the heart and brain that may lead to a heart attack or stroke.

Diet is a modifiable cardiovascular and cerebral risk factor, meaning that when one's diet is heart-healthy then risks decrease.

Heart Disease: Terms to Know

Angina pectoris is a condition that is marked by severe chest pain and/or discomfort that often spreads to the neck and shoulders. It is frequently due to insufficient blood flow to the heart muscle that results from obstruction or spasm within the coronary arteries.

Ischemic heart disease is characterized by reduced blood supply to the heart due to narrowing of the arteries and less blood and oxygen delivered to the heart muscle. Ischemic heart disease is the most common cause of death in Western countries.

A **heart attack** is sometimes referred to as a myocardial infarction. It occurs when one or more regions of the heart have severe oxygen debt. This may be the result of blood clots that block the flow of blood through coronary arteries to the heart muscle.

The average age for a first heart attack in the U.S. is 65 years old. However, it has been estimated that four to ten percent of U.S. men aged 45 years or younger experience a heart attack and that **atherosclerosis**, or hardening of the arteries, starts in youth.

The risk factors for heart attacks include age, autoimmune disease, diabetes, family history of heart attack, high blood cholesterol or triglyceride levels, high blood pressure, illegal drug use, lack of physical activity, obesity, preeclampsia, stress, and tobacco use.

A **stroke** occurs when the blood supply to the brain is either blocked or a blood vessel within the brain ruptures. In either instance, brain tissue may die and brain health may be compromised or this may lead to death.

A stroke is considered a medical emergency and treatment is needed very quickly. It is the fifth leading cause of death in the U.S.

There are three main types of strokes: ischemic, hemorrhagic, and mini-strokes (transient ischemic attacks). An ischemic stroke is the result of an obstruction within a blood vessel that supplies blood to the brain. A hemorrhagic stroke occurs when a weakened blood vessel ruptures. A transient ischemic attack occurs when the blood supply to the brain is briefly blocked.

In comparison to a heart attack, the risks for having a stroke of any kind doubles each decade after the age of 55 years, though strokes may occur at any age. In fact, almost one-fourth of strokes occur in the U.S. to people under 65 years of age, especially when they have risks factors that include alcohol consumption, gender (women more than men), oral contraceptive use, post-menopausal hormone therapy, race (Asian, African American, and Hispanic in particular), tobacco use, and past heart attacks or strokes.

Coconut Consumption

Let's examine populations around the world with higher coconut consumption than in the United States and see what we can learn from these populations in terms of coconuts in history, cardiovascular disease, health, and longevity. There is a rich history of coconut consumption for nutrition, health, and well-being in countries around the world where coconuts have thrived for centuries.

Kitavans

The Kitavans who reside on the Trobriand Islands, Papua New Guinea, consume a significant amount of coconuts and reportedly exhibit excellent health. A subsistence lifestyle is still followed by Kitavan islanders. Dietary staples include coconuts, fish, fruit, and tubers. Stroke and ischemic heart disease appear to be absent in this population.

In a study that was conducted in the 1990s, when the frequencies of aphasia, exertion-related chest pain, hemiparesis, sudden imbalance, or spontaneous sudden death were assessed in Kitavan islanders, no cases that corresponded to angina pectoris, stroke, or sudden death were reported.

Pukapukas and Tokelauans

The Pukapukas and Tokelauans from the southern Pacific Ocean are also sizeable coconut consumers. Over 60 percent of their daily calorie intake is from coconuts and they reportedly have the largest saturated fat consumptions in the world with the absence of heart disease.

The diets of the atoll dwellers near the equator from both Pukapuka and Tokelau are high in saturated fats but low in dietary cholesterol and sucrose. Coconut is the main source of energy for both of these population groups.

A study from the 1980s showed that Tokelauan people acquire a higher percentage of energy (calories) from coconuts than people from Pukapuka: 63 percent compared to 34 percent, with a higher intake of saturated fat. In comparison, the serum cholesterol levels in Tokelauans were 35 to 40 milligrams higher than in the Pukapukans.

The difference in serum cholesterol level was attributed to the higher saturated fat intake of the Tokelauans due to the high lauric acid (12:0) and myristic acid (14:0) content. In both populations vascular disease is uncommon and there is no evidence of high saturated fat intake having detrimental health effects in these populations.

In order to truly understand the cardiovascular and brain benefits that the coconut consumption provides for these indigenous populations, an examination of heart-healthy diets and the heart disease process is warranted.

Heart-Healthy Diets

In the 1950s the connection between saturated fats and cardiovascular disease was widely circulated throughout the U.S. and lower-fat diets were developed and promoted for cardiovascular disease protection.

Then, in the 1980s, the first set of U.S. Dietary Guidelines were established by the Departments of Agriculture (USDA) and Health and Human Services (HHS). These new guidelines recommended that all Americans avoid eating too much total fat, saturated fat, and cholesterol. The U.S. Dietary Guidelines were revised every five years and Americans were advised, among other recommendations, to continue to follow lower total fat, lower saturated fat, and lower cholesterol diets.

In the 1990s, while the recommendtions for dietary fats generally remained the same, the recommendation for dietary carbohydrates instructed that Americans, "Choose a diet with plenty of grain products with vegetables and fruits."

Some dietary fats in foods and beverages were replaced by carbohydrates and often by sugars. This led to recommendations to "Choose a variety of grains daily, especially whole grains; choose a variety of fruits and vegetables daily; choose beverages and foods to moderate your intake of sugars and choose a diet that is low in saturated fat and cholesterol and moderate in total fat."

These guidelines may have been misinterpreted by the U.S. public, who thought that carbohydrates could be freely consumed—as long as they were healthy whole grains. As a result, higher carbohydrate diets were popular at the close of the twentieth century.

Concurrently, fats such as butter and lard were replaced with manufactured margarines, most with trans fatty acids (which have a relationship to increased cardiovascular disease).

In conjunction with these 50 years of U.S. Dietary Guidelines, after peaking around 1968, the age-adjusted rates from cardiovascular disease were reduced in half by the year 2000. Two factors mainly contributed to this reduction: substantial decreases in elevated total cholesterol, high blood pressure, and tobacco use (major cardiovascular risk factors) and new treatments for established cardiovascular disease. These include angioplasty, bypass grafting, stents, enzyme (ACE) inhibitors, statins, and thrombolysis. Deaths from strokes also declined over this same period of time. However, the prevalence of both diabetes and obesity has alarmingly increased.

Today saturated fatty acids, such as those that are found in coconuts, are looked at differently than before for their relationship to cardiovascular disease prevention and health enhancement. An examination of differences among fatty acids helps to sort out which are healthier by today's standards and why and how coconuts fit into the total cardiovascular picture.

A Fat Primer

Cholesterol is a fatty substance that occurs in only animal foods and products. Organ meats, high-fat dairy products, and some seafood are higher in cholesterol. Cholesterol is a contributing factor in cardiovascular disease, but non-dietary factors may also raise cholesterol and increase one's predisposition to cardiovascular disease. There are primarily three types of serum cholesterol that are considered predictive of cardiovascular disease:

- **High-density lipoproteins (HDL-cholesterol):** known as the "good cholesterol" because it proportionally carries more protein than cholesterol from the blood to the liver to be recycled or for disposal. HDL-cholesterol is affected more by a person's activity level, gender, obesity, and smoking than diet.

- **Low-density lipoproteins (LDL-cholesterol):** known as the "bad" cholesterol because it mainly carries cholesterol and is associated with increased risk of cardiovascular disease.

Dietary cholesterol, saturated fat intake, total dietary fat, trans fat, and excess calories are some of the dietary factors that are thought to increase LDL-cholesterol, in addition to the lifestyle factors of obesity, inactivity, and smoking.

- **Very-low density lipoproteins (VLDL-cholesterol):** is also known as the "bad" cholesterol. It is produced by the liver and released into the blood to supply triglycerides (see below). High VLDL-cholesterol is associated with plaque development on artery walls that narrows and restricts blood flow. By exercising and losing weight and avoiding sugary foods and alcohol, the body's serum (blood) triglyceride level may be decreased.

Lipids are a family of fats, oils, phospholipids (such as lecithin), and sterols (such as cholesterol) that are found in plants or animals, or sometimes both. Cholesterol is only found in animals. Sometimes lipids are inadvertently called "fats," but they really contain many more substances.

Fats are lipids that are solid at room temperature. Animal fats include bacon, butter, lard, the skin of poultry, and the fat that runs though meats. Generally speaking, the more solid the fat, the more difficult it is for the body to break down.

Hydrogenation is the process of converting unsaturated fatty acids into saturated fatty acids through a method of trans bonds. As a result, these fatty acids are very difficult for the body to break down and they may contribute to cardiovascular disease.

Partially hydrogenated means that some fatty acid bonds are "partially" hydrogenated or filled, which makes them somewhat easier to digest. Instead of risky trans bonds, partially-hydrogenated fatty acids have different linkages, or cis bonds. Some frostings, margarines, and shortenings are all foods with partially-hydrogenated products.

Phospholipids are similar in structure to triglycerides, but they also contain phosphorus. They also provide the major structural components of membranes. Lecithin is a phospholipid that naturally occurs in animals, plants (in soybeans), and eggs, among other foods. Lecithin is used as an emulsifier by the body and by the food industry.

Phospholipids in the form of lecithin are found in egg yolks, liver, peanuts, and wheat germ. They are also found in milk, soybeans, and lightly cooked meats. The human body can synthesize lecithin if there is sufficient choline in the diet from dairy foods, eggs, fish, meats, poultry, pasta, or rice.

Oils are lipids that are liquid at room temperature. They include such oils as avocado, canola, coconut, olive, peanut, and safflower. While coconut oil is semisolid at room temperature (due to its saturated fat), it is referred to as an oil.

Essential fatty acids must be consumed by the body and are essential for its normal functioning. There are two essential fatty acids (EFAs) that cannot be synthesized by the body: *linoleic* and *alpha-linolenic*. These must be obtained by food. These two essential fatty acids are used to build omega-3 fatty acid and omega-6 fatty acid, both important in normal body functioning (see below). Salmon, herring, mackerel, hemp, flax, walnuts, almonds, dark leafy green vegetables, olive oil, whole grains, and eggs are good sources of essential fatty acids. The human body can generally produce non-essential fatty acids from other substances.

Monounsaturated fatty acids are unsaturated at one point in their fatty acid structures where they can be broken down, so they are considered to be healthier. Monounsaturated fatty acids are primarily found in canola, olive, peanut, safflower, and sesame oils, as well as in avocadoes, peanut butter, and many nuts and seeds.

Omega-3 fatty acids (Omega 3's) are essential fatty acids that are associated with heart health and disease prevention. Omega 3-fatty acids may help to decrease cholesterol and platelet clotting and inflammatory and immune reactions. They are primarily found in chia seeds, fatty fish, fish oil, flaxseeds and flax oil, seafood, soybeans, spinach, and walnuts.

Omega-6 fatty acids (Omega 6's) are essential fatty acids that have a number of body functions. Along with omega-3 fatty acids, omega 6-fatty acids play a critical role in healthly brain function and normal growth and development. While omega-3 fatty acids tend to decrease inflammation, omega-6 fatty acids tend to promote it and may play a role in complex regional pain syndrome.

Polyunsaturated fatty acids have many bonds that are unsaturated so they can easily be broken down by the body. This is not always a good feature since this may lead to instability, particularly when they are located within cellular membranes. Foods with higher amounts of polyunsaturated fats include corn oil, fish (such as albacore tuna, herring, mackerel, salmon, and trout), safflower, soybean, and sunflower oils, and walnuts.

Saturated fatty acids are fully saturated, which means that their bonds are filled. This feature normally makes them difficult for the body to break down, or metabolize. Generally saturated fatty acids are found in animal foods and beverages; however, coconut and palm oils are noted for their saturated fatty acid content.

Sterols (also known as steroid alcohols) are a type of steroid that occur naturally in animals (such as cholesterol), fungi, and plants (such as phytosterols that are found in soy foods). Cholesterol is vital to animal cell membranes. It also functions as a precursor for making fat-soluble vitamins (vitamins A, D, E, and K) and steroid hormones. Dietary cholesterol also contributes to increased risks of cardiovascular disease when it is elevated in blood cholesterol. Phytosterols, or plant sterols, have been shown to block cholesterol absorption and help to reduce serum cholesterol.

Trans fatty acids are of two kinds: naturally-occurring and artificial. Naturally occurring trans fatty acids are formed in the gut of some animals and foods (such as milk and meat products). Artificial trans fatty acids are created during industrial processes that add hydrogen using trans bonds to solidify liquid vegetable oils. Trans fatty acids were used in the U.S. because they were easy to use, inexpensive, and long-lasting. However, trans fatty acids raise LDL ("bad") cholesterol and lower HLD ("good") cholesterol and increase the risks of developing cardiovascular disease, type 2 diabetes, and stroke. In 2015, the U.S. Food and Drug Administration required that synthetic trans fatty acids be phased out of all foods within three years.

Triglycerides are the main form of lipids that are found in the body and in foods and beverages. Triglycerides are composed of a backbone molecule of glycerol (a sugar alcohol) and three fatty acids that can be saturated, unsaturated, or both. High serum triglycerides are associated with increased risks of cardiovascular disease and strokes and conditions such as hypothyroidism, kidney or liver ailments, metabolic syndrome, obesity, and uncontrolled type 2 diabetes may also be associated with elevated serum triglycerides. By eliminating alcohol, limiting dietary cholesterol and some saturated fats, losing weight, reducing simple carbohydrates, and restricting trans fatty acid intake, the serum triglyceride level may decrease.

Unsaturated fatty acids have one or more of their bonds that are unsaturated, which means that they are not filled. This is generally a good feature since the body can break down and metabolize unsaturated fatty acids more easily. Almonds, avocadoes, flaxseeds, hazelnuts, macadamia nuts, peanut butter, salmon, sardines, seeds, vegetable oils, and walnuts contain unsaturated fatty acids.

How Coconuts Compare

The cardiovascular benefits of coconuts are revealed when each of these fats and oils are compared to the types of fats in coconuts. A comparison of raw coconuts, coconut oil, and coconut water follow:

Cholesterol (high-density or HDL, low-density or LDL, and very-low density or VLDL)

- Since coconuts are from the plant kingdom, raw coconut, coconut oil, and coconut water do not contain any form of cholesterol.

Fats

- One cup of raw coconut contains 26.8 grams of total fat that are mostly saturated (23.8 grams compared to 26.8 grams total).
- One cup of coconut oil contains 218 grams of total fat that are mostly saturated (189 grams compared to 218 grams total).
- One cup of coconut water contains virtually no total fat (0.5 grams per one cup) or saturated fat (0.4 grams per one cup).

Hydrogenation

- Since hydrogenation is a process of converting unsaturated fatty acids into more saturated fatty acids, coconuts in their raw state do not contain hydrogenated fatty acids.

- Hydrogenated coconut oil

−A small portion of the unsaturated fatty acids in coconut oil may be hydrogenated in order to keep the coconut oil solid at higher temperature. Outside of the U.S., hydrogenated and partially hydrogenated coconut oils may be used in the confection industry in baked goods or candies in tropical climates.

Lecithin

- Lecithin may be combined with coconut as an emulsifier; however, coconut oil with its higher concentration of saturated fatty acids is already semi- to fully solidified.

Lipids

- Because lipids are fats and oils, the fats that are found in coconuts are considered lipids.
- Since there is virtually not any total fat or saturated fat in one cup of coconut water, it does not significantly contain any lipids.

Oils

- Coconut oil can be extracted from raw coconut meat. The extraction rate may be as much as 75 percent.
- Coconut oil is in a semisolid state at room temperature.

Essential fatty acids

- Omega-3 fatty acids

−There are no omega-3 fatty acids in one cup of raw, shredded coconut, one cup of coconut oil, or one cup of coconut water.

- Omega 6 fatty acids

−There are 293 milligrams of omega-6 fatty acids in one cup of raw, shredded coconut.

−There are 3,923 milligrams of omega-6 fatty acids in one cup of coconut oil.

−There are 4.8 milligrams of omega-6 fatty acids in one cup of coconut water.

Phospholipids

- Phospholipids may be present in coconut endosperm and coconut oil.

Sterols

- Phytosterols

–There are 37.6 milligrams of phytosterols in one cup of raw, shredded coconut.

–There are 187 milligrams of phytosterols in one cup of coconut oil.

–There are virtually no phytosterols in one cup of coconut water.

Saturated fatty acids

- The saturated fatty acids in coconuts are lauric (C12), myristic (C14), and stearic (C17).

–There are 23.8 grams of saturated fatty acids in one cup of raw, shredded coconut.

–There are 189 grams of saturated fatty acids in one cup of coconut oil.

–There are virtually no saturated fatty acids (0.4 grams) in one cup of coconut water.

Trans fats

–There are no trans fatty acids in raw coconut, coconut oil, or coconut water.

Triglycerides

- The saturated fatty acids in coconuts are medium-chain triglycerides that are extracted from coconut meat into coconut oil.

Unsaturated fatty acids

- Monounsaturated fatty acids

–There are 1.1 grams of monounsaturated fatty acids in one cup of raw, shredded coconut.

–There are 12.6 grams of monounsaturated fatty acids in one cup of coconut oil.

–There are virtually zero grams of monounsaturated fatty acids in one cup of coconut water.

- Polyunsaturated fatty acids

–There are 0.3 grams of polyunsaturated fatty acids in one cup of raw, shredded coconut.

–There are 3.9 grams of polyunsaturated fatty acids in one cup of coconut oil.

–There are virtually zero grams of polyunsaturated fatty acids in one cup of coconut water.

Fat to Fight Fat?

Coconut, even with its high total fat and saturated fat content, has actually been shown to help aid weight loss. In this section the various ways that the coconut may contribute to the fight against body fat are discussed in depth.

For example, the medium-chain triglycerides (MCTs) in coconut oil may actually help to *burn* body fat and boost metabolism. Coconut has been thought to help metabolize abdominal fat. This is especially encouraging since excess abdominal fat is a known risk for heart disease and diabetes, among other health conditions.

While the fat in coconut may help weight loss, it is important to recognize that it has the same number of calories per gram (9 calories/gram) as any other dietary fat, so coconut still should be consumed in moderation. Yet there are still smart ways to incorporate coconuts into healthy diets without being too excessive.

Using coconut oil for cooking instead of canola or olive oil may result in a *thermogenic effect* within the body, which means that it may help increase metabolism and in turn, increase energy. While this amount of energy may be low, it still adds up. Even more, eating coconut can lead to feelings of satiety and reduced appetite. This may result in fewer calories consumed overall.

This section also touches on some of the properties of coconut water. Unlike coconut oil, coconut water is virtually fat-free; it's also low in calories and contains nutrients such as vitamin C and potassium. Coconut water can be used for hydration, especially in place of sugary sports drinks or soft drinks, which is very helpful when people try to lose weight healthfully.

Energy

Energy is the ability to do work. It is measured in *calories*. A *calorie* is a unit of energy that often refers to the calorie content of foods and beverages. A calorie produces heat—in fact, a calorie is defined as the amount of heat that is required to raise the temperature of one kilogram of water one degree Centigrade. In the human body, the process of burning calories to produce energy (and heat) increases one's metabolism.

Metabolism is the sum of all of the chemical and physical processes by which energy is created and made available for life. There are many factors that are involved in metabolism and whether a person metabolizes or burns food and beverages quickly or slowly.

Factors that lower metabolism include age, fasting, hormones, sleep, and starvation. As a person ages their metabolism slows down, as it does with severe dieting. Hormones produced by the thyroid gland may either increase or decrease metabolism. And too much sleep means that the body is less active; thus, metabolism may be slower while a person is sleeping.

Energy-producing nutrients (carbohydrates, lipids, and proteins) also affect metabolism; their effect depends on whether or not they are consumed and in what forms and combinations. As a type of lipid with very distinctive properties, coconut oil and its metabolism into energy can actually contribute to fat loss if used properly. However, do keep in mind that consuming any nutrient in excess may lead to weight gain.

Calories from Nutrients

The energy-producing nutrients each provide a distinct number of calories per gram when they are metabolized. Carbohydrates (sugars and starches) and proteins each yield about four calories per gram (4 calories/gram) while fats and oils each yield about nine calories per gram (9 calories/gram)—about twice as much as carbohydrates or proteins. This is why fats and oils are considered to be so caloric and "fattening."

Weight Loss Strategies

In order to lose weight, a person needs to eat and drink less, exercise more, or some combination of both of these approaches. There must be a caloric deficit for weight loss. Conversely, if more calories are consumed rather than expended, weight will probably be gained.

There are many dietary strategies to accomplish weight loss including gluten-free, high-carbohydrate, high-fiber, high-protein, low-calorie, low-carbohydrate, and countless other approaches, programs, products, and potions. The use of coconut oil with its medium-chain triglycerides may have advantages. To understand how the medium-chain triglycerides that are found in coconuts contribute to weight loss, one must first understand more about fat digestion and metabolism.

Fat Digestion and Metabolism

The digestion of *lipids* (fats and oils) first begins in the mouth, where it is mostly physical. The teeth tear apart fat from other foods and the mouth begins to melt some of the fats. Then the residue passes through the esophagus into the stomach where it is mixed with stomach acid and water and some fat is broken down by the acid.

The process of breaking down lipids into its components of *triglycerides* with their fatty acids and *glycerol* is slow. Since lipids have so many calories per gram, they tend to stay around the stomach longer and may lead to satiety or fullness.

Once the triglyceride remnants pass into the small intestine, the smallest fatty acids (or short-chain fatty acids) and glycerol are able to pass through the intestinal wall into the bloodstream where they are transported to the liver, stored, or converted into other substances.

Longer-chain triglycerides (LCTs) are broken down by bile in the small intestine. *Bile* is an emulsifier that is made in the liver and stored in the gallbladder. It mixes with lipids along with watery digestive secretions and readies the lipids for additional breakdown by intestinal enzymes.

Medium-chain triglycerides (MCTs) have a different fate. For one, they are more easily metabolized than longer-chain triglycerides and may help to burn calories and support weight loss. Medium-chain triglycerides passively diffuse from the gastrointestinal (GI) tract through the portal system to the liver without needing to be broken down, nor the necessity for bile salts for their digestion.

Medium-chain triglycerides either do not affect total cholesterol nor raise HDL cholesterol and they may also improve the ratio of LDL to HDL-cholesterol (LDL:HDL). Medium-chain triglycerides are also thought to promote quick fat oxidation. This is why some endurance athletes and bodybuilders seem to favor MCTs. However, claims and conclusive results about their effectiveness seem to be mixed.

Besides all of these applications, medium-chain triglycerides may also have the potential to promote weight loss by increasing the metabolism. To fully comprehend why, an understanding of ketosis, particularly for weight loss, is essential.

Ketosis for Weight Loss

Ketogenic diets have been used for fast (and in some cases continued) weight loss for centuries. Ketogenic diets are also said to show promising results for epilepsy and neurodegenerative disorders, such as Alzheimer's and Parkinson's diseases, as discussed in the next section.

First of all, let's answer a basic question: what is ketosis? Normally the human body uses glucose for energy. *Glucose* is a simple sugar that is the building block of carbohydrates (sugars and starches). When the human body metabolizes carbohydrates, they break down into glucose that can be used for energy or stored for future energy needs. Consuming too many carbohydrates may lead to excess calories and this may progress to obesity.

When glucose is not available for energy, the human body turns to an "intermediary" state for its energy needs. This is called *ketosis* when the human body produces *ketones*, by-products that form when the body burns stored fat for energy.

Some higher-protein and fat and lower-carbohydrate diets rely upon ketosis for fat loss and decreased body weight. Long term and uncontrolled, ketosis may develop into a state called *acidosis* that may lead to coma or even death. Ketosis may also occur during prolonged starvation.

To comprehend why ketogenic diets work and how medium-chain triglycerides interplay, one must first understand the process of *lipolysis*. Like glycolysis and the breakdown of carbohydrates to produce energy, lipolysis is the breakdown of lipids to produce energy.

When the body needs energy and it calls upon its stored carbohydrates, it may also call upon its lipid stores within its cells—particularly when carbohydrates "run out" as in prolonged starvation or long-term exercise. Enzymes break down stored lipids; then fatty acids and glycerol (from triglycerides) are released into the blood stream. When these fatty acids reach the muscle cells they enter *mitochondria* of the cell, or the cell's powerhouse and energy is released.

As mentioned, lipids have the capacity of supplying twice the amount of energy than carbohydrates or protein (9 calories/gram compared to 4 calories/gram respectively). This is the reason why fats and oils are so calorie dense and also why ketogenic diets may lead to faster weight loss than simple calorie reduction or some other carbohydrate-based diets.

Medium-Chain Triglycerides and Weight Loss

The saturated fatty acids that are found in coconut oil with their medium-chain triglycerides are shorter in length than longer-chain triglycerides and they are more water-soluble than some other oils. The body processes these medium-chain triglycerides differently than longer-chain triglycerides. Longer-chain triglycerides must be mixed with a substance called *bile* that is produced by the gallbladder and then assaulted by pancreatic enzymes. Medium-chain triglycerides do not require bile or pancreatic enzymes, so once they reach the small intestine they diffuse through its membrane into the bloodstream where they are more directly routed to the liver, converted into ketones, and metabolized as fuel.

The liver also releases ketones back into the bloodstream where they are carried throughout the body. In this manner, ketones may be used for energy by the brain. This is significant since the brain has a protective barrier called the "*blood-brain barrier*" that these ketones can cross.

As a result, ketones have fewer tendencies to be deposited in fat stores and more opportunities to be metabolized by the body for various purposes. Additionally, medium-chain triglycerides provide antioxidant, anti-inflammatory, and even antimicrobial functions inside the gut where they fight potentially harmful bacteria, fungi, parasites, and viruses.

Medium-Chain Triglyceride Structures

Medium-chain triglycerides are distinguished by their carbon lengths:

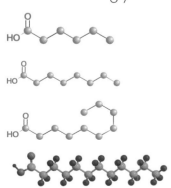

- Caproic acid (C6) — 6 carbons
- Caprylic acid (C8) — 8 carbons
- Capric acid (C10) — 10 carbons
- Lauric acid (C12) — 12 carbons

In general, the fewer number of carbon molecules, the shorter the chain and the easier and more efficiently that medium-chain triglycerides will be converted into ketones for energy with less reactive oxygen that may be damaging to the body.

Coconut oil has a mixture of medium-chain fatty acids, with about 40 percent of it as *lauric acid*.

Some studies suggest that coconut oil with its medium-chain triglycerides and particularly lauric acid may help to reduce one's midsection, but that it may not contribute to improved body mass index (BMI) or significant weight loss.

The ketones also help to control the hormone *ghrelin*, the "hunger" hormone that is produced in the gastrointestinal tract. Ghrelin functions within the central nervous system to increase appetite. *Leptin*, the appetite "suppressor" hormone, controls ketone utilization in the neurons of the *hypothalamus* (the hunger control center of the brain) that helps to decreases one's appetite.

Medium-Chain Triglycerides and Weight

It is known that medium-chain triglycerides may boost *thermogenesis* (heat production) and fat oxidation for energy and that both of these processes may suppress the accumulation of body fat. Another way that medium-chain triglycerides may help with weight loss efforts is by helping with appetite reduction.

Consuming foods and beverages with fat may turn off a desire to eat more calories because fats provide such concentrated calories. Even a little bit of fat that is consumed at meals has staying power.

There is also some speculation that medium-chain triglycerides act on other hormones that may include cholecystokinin, gastric inhibitory peptide, neurotensin, pancreatic polypeptide, and/or peptide YY. The exact modes of operation are unknown. However, it is speculated that these actions may help to decrease appetite and increase satiety.

Use of Medium-Chain Triglycerides by Endurance Athletes

Coconut oil has been valued for its relationship to weight and energy expenditure among endurance athletes. Coconut oil supplies energy; it has the capability to support fitness since it may boost energy, enhance performance, and support endurance. Also, coconut oil may improve the digestion and absorption of nutrients—especially those that require a little fat for their digestion and assimilation. And coconut oil is thought to assist in blood sugar utilization and insulin secretion. Suggested dosages vary.

Since medium-chain triglycerides are digested easier than longer triglycerides, the thought is that they may help to increase energy metabolism during both medium and high-intensity exercise. Medium-chain triglycerides helped to reduce the body's reliance on carbohydrates as a fuel for exercise and decrease the amount of *lactate* that is produced during exercise that may lead to increased endurance.

High lactate levels increase the acidity of the muscle cells and disrupt other metabolites. Metabolic pathways perform poorly within this type of acidic environment. This is a natural defense mechanism of the body to help prevent permanent damage during extreme exertion by slowing down key systems to help maintain muscle contraction. Then oxygen becomes more available and metabolism can continue for energy and recovery from strenuous exercise. Medium-chain triglycerides may help to prevent or reduce these processes from occurring.

Disadvantages of Medium-Chain Triglycerides

Medium-chain triglycerides may induce *ketogenesis* (the release of ketones by the body when fats are broken down for energy) and *metabolic acidosis* (a condition that occurs when the body produces too much acid or when the kidneys are not removing enough acid from the body). Under certain conditions, ketogenesis may be desirable. However, metabolic acidosis may lead to shock or death.

If medium-chain triglycerides are consumed in high quantities they may provoke gastrointestinal side effects including loose stools.

Coconut Water, Weight, and Sports Performance

Coconut water is virtually fat-free and low in calories with some nutrients including vitamin C and potassium. It can be used for hydration to replace sugary sports drinks or soft drinks, which is important for people who try to lose weight and for athletes to remain healthy while they train and compete.

A comparison of water, coconut water, sports drinks, and soft drinks shows the advantages and disadvantages of these fluids.

Water is a fluid of choice for everyday consumption and for everyday exercise. For exercise that is greater in duration or intensity, or for exercisers who heavily

perspire, some fluids may be better for hydration. Some brands of sports drinks and coconut water with potassium may be effective for muscle cramps.

Coconut water is also a natural way to add potassium, hydrate, and reduce sodium in everyday diets that rely upon sports drinks as regular beverages. This is because coconut water has fewer calories and less sodium than some sports drinks. Also, many Americans do not obtain enough potassium in their daily diets from dairy products or fresh fruits and vegetables, so coconut water may help to provide some needed nutrients.

When a person exercises strenuously in excess of three hours (as the time expended in a marathon) and they sweat heavily, they will probably need easily absorbable carbohydrates for energy (such as the sugar that is found in sports drinks), as well as electrolytes such as potassium and sodium. This is why unsweetened coconut water may be limiting in its use for hydration and sports drinks may be more satisfying and hydrating under extreme conditions.

An option during hot weather events or practice might be to mix coconut water with a little salt. Coconut water with added sodium may be better tolerated than some sports drinks post-exercise and may not cause as much fullness, stomach upset, or nausea as some report.

One concern about the use of coconut water for exercise is whether or not the amount of coconut water consumed replenishes what the body loses during exercise. A simple test is the color of urine that should be clear. It is important to note that taste matters when it comes to replacing body fluids. So if athletes enjoy the taste of coconut water and are able to consume coconut water in adequate amounts, then it might be more efficient for hydration than water.

Comparison of Calories

A comparison of calories, carbohydrates, fat, and protein in water, coconut water, sports drinks, fruit juice, and cola follow:

- **Tap and non-mineralized water (per 1 fluid ounce)**

—Zero calories, zero carbohydrates, zero fat, zero protein

- **Unflavored coconut water (per 1 fluid ounce)**

—5.45 calories, 1.3 grams of sugar, 61 milligrams of potassium, 5.45 milligrams of sodium

- **Sports drinks (per 1 fluid ounce)**

—6.25 calories, 1.75 grams of sugar, 3.75 milligrams of potassium, 13.75 milligrams of sodium

- **Fruit juice (per 1 fluid ounce)**

—3.9 calories, 2.6 grams of sugar, 62 milligrams of potassium, 0.3 milligrams of sodium

- **Cola (per 1 fluid ounce)**

—11.3 calories, 2.7 grams of sugar, 0.6 milligrams of potassium, 1.2 milligrams of sodium

Coconut Water for Weight Management

Consuming a liquid before a meal may cause one to eat less. This is the theory behind a "slow-up liquid." A slow-up liquid can be sipped before a meal, with or without ice and a flavorful herb or fruit slice to boost the flavor.

Try coconut water with a sprig of basil, mint, or parsley or a spear of cucumber, jicama, or fennel. The calories are fairly negligible. It may seem and look like a cocktail, to refresh, and help to mitigate the ravishing feeling some may feel right before a meal.

To add the flavor of coconut water to beverages without packing on calories, try freezing coconut water in an ice cube tray, then adding a few cubes to perk up the flavor of ordinary water. The taste will be appealing without overwhelming and may be enjoyed by those who do not enjoy a stronger coconut taste.

For those people who can afford more calories, but prefer to avoid commercial sports drinks, mixing one-half coconut water with one-half fruit juice provides a fuller fruit taste with half of the calories than a full glass of fruit juice. Plus, there is more potassium along with varying amounts of choline, folate, vitamins A, C, calcium, magnesium, and phosphorus depending upon the type of coconut water that is used.

Coconut water also has many applications in cooking and baking. It can be used in some recipes in place of a few teaspoons of water, but keep in mind its sweet taste. In some recipes, coconut water will enhance the sweetness. In other recipes, it may quash the bitterness. And in other recipes, the sweet taste might be tamed by a touch of salt, which may even taste tropical.

Speaking of the tropics, coconut water may be an excellent addition to some Asian or Caribbean recipes because of its natural affinity to the tropical fruits and vegetables in these recipes, but it may also lend a sweet and tasty touch to marinades, sauces, or even soups. It contributes the coconutty taste without the heaviness of coconut oil. In addition, coconut water can be used in some recipes along with coconut oil and even grated coconut.

Use Your Brain

Some of the more interesting research that surrounds the healing power of coconut involves the possible links between coconut oil and decreased dementia and epilepsy. Much of this research surrounds the ketogenic diet: a diet that is higher in fat and lower in carbohydrates than is generally recommended, which forces the body to burn fat for fuel.

This section describes the ketogenic diet in detail, how coconuts may fit into ketogenic diets, and why ketogenic diets may be beneficial for certain disorders. For example, the ketogenic diet has been shown to be effective at reducing seizures in children with epilepsy. This discovery has led scientists to theorize that coconut oil may be useful for people with Alzheimer's disease and dementia.

This section also describes methods by which the coconut may prevent general aging in the brain and help improve memory.

Ketogenic Diets

The *ketogenic diet* has been used for a variety of disorders since the 1920s when it was designed by Dr. Russell Wilder at the Mayo Clinic in Rochester, Minnesota, as a treatment for epilepsy. After anti-seizure medications came to the forefront in the 1940s, the ketogenic diet somewhat lost its appeal. While low-carbohydrate, ketogenic-like diets have a long history, they became popularized in the U.S. in the late 1960s to early 1970s when the Stillman and Atkins diets first became popular. Their premise is pretty much the same: the process of ketosis could effectively be employed for weight loss.

The premise of ketogenic diets is that with higher-fat, adequate protein, and low carbohydrates the body is forced into "burning" its stored body fats for energy rather than carbohydrates, the preferred energy source.

Under "normal" conditions, the human body digests and metabolizes the carbohydrates that are contained in foods and beverages and converts them into glucose for energy. Glucose is then transported to many various body cells and also fuels the brain.

When there are few carbohydrates available, then the liver is capable of converting fat into fatty acids and *ketone bodies*, molecules that can also be oxidized for energy. These ketone bodies can then pass through the *brain barrier* (a highly selective semipermeable membrane that protects the brain) and replace glucose as a source of energy. When ketones are elevated in the blood it may result in a state of ketosis that has both advantages and disadvantages.

The advantages of ketosis are that it can be used as a medical intervention for diabetes or epilepsy, endurance exercise, or significant weight loss. Ketogenic diets may reduce hunger and lower overall food intake that may be particularly effective in treating the severely obese.

The disadvantages of ketosis are that it may alter blood lipid levels (both positively and negatively), lead to fatigue, stress the kidneys, trigger micronutrient deficiencies, and instigate ketoacidosis, which could be fatal at extremes for diabetics.

Epilepsy

Fasting to treat disease has been a topic of discourse since ancient Greek and Indian physicians utilized it in their practices. An early treatise in the Hippocratic Corpus "On the Sacred Disease" (supposedly written in 400 BC) describes how dietary alterations may affect epilepsy management—particularly in the abstinence of food and drink.

The first study that examined fasting as an epileptic treatment occurred in France in 1911. Patients were placed on a low-calorie vegetarian diet instead of receiving the treatment of the time, potassium bromide. While potassium bromide curbed the epileptic seizures it also reportedly lowered the mental capacity of the study subjects. The diet plus fasting led to improved mental capabilities.

Three main ketone bodies were at the forefront of diabetes research in the 1920s. When otherwise healthy people were starved or consumed a very low-carbohydrate, high-fat diet, then a higher level of these ketone bodies were produced. This higher level of serum (blood) ketones is referred to as *ketonemia*.

Subsequently, anticonvulsant treatments were used to treat epilepsy. Still, epileptic control was not fully achieved. The ketogenic diet was later reintroduced as a therapy to treat children with epilepsy. After an 18 to 25 day fast some children became seizure free; others showed some improvement, while still other children demonstrated the most success on what was referred to as a "water diet." The idea of fasting subsequently became a conventional treatment for epilepsy along with abidying by dietary carbohydrate restriction.

Weight Loss

The use of ketone bodies has also been used for dieting nearly as long as they have been used to treat epilepsy.

In the 1920s, an endocrinologist detected that ketone bodies, produced by the liver as a side effect of fasting or a diet that was high in fat and low in carbohydrates, were effective for weight loss. This type of dietary approach became known as the low-carbohydrate/ketogenic diet. It has had many guises throughout the years. Some of the books that recommended it included the 1958 book *Eat Fat and Grow Slim* by Dr. Richard Mackarness; the 1967 book *The Doctor's Quick Weight Loss Diet* by Dr. Irwin Stillman; and the 1972 book *Dr. Atkins Diet Revolution*.

Ketogenic diets may contain 70 percent or more daily calories from fat; 5 to 10 percent of total daily calories from carbohydrates (about 20 to 50 grams daily); and the remaining calories from proteins. Some more conservative dietary approaches are higher in carbohydrates (45 to 65 percent of total calories), lower in fat (20 to 35 percent of total calories), and moderate in protein (10 to 35 percent of total calories). Since 1980 and every five years thereafter, the U.S. Dietary Guidelines for Americans were published with renditions of these dietary approaches, which are now less specific in nature.

In order to adhere to a ketogenic diet with these nutritional parameters, one needs to greatly reduce or eliminate carbohydrate-containing foods such as dairy products, fruit, grains, legumes, and starchy vegetables. High protein foods such as eggs, fish, shellfish, meats, and poultry are the mainstays with very-low calorie vegetables as fillers. Fats as butter and oils are also featured. Alcohol is not included.

Some studies show that weight loss may be greater on a ketogenic diet than on a diet that is simply reduced in calories. Other studies indicate that the average weight loss after one year on a ketogenic diet may be greater than a low-fat diet and that serum triglycerides and blood pressure may decrease while high-density lipoproteins (HDL, or "good" cholesterol) may increase. Even if weight loss stabilizes, after a few years high-density lipoproteins may continue to improve on a ketogenic diet. Like many diets, weight loss slows as a result of the body acclimating to its dietary parameters.

Positive Aspects of Ketogenic Diets

Meat has been a significant foundation of the U.S. diet, especially post World War II when it has been associated with prosperity. Aside from economics, meat is very satisfying with its fat content and robust taste. It has a very primeval appeal since its protein content is essential for life. Both prehistoric and modern people have sought meat for its amino acids—the staff of life. The ketogenic diet, with its high reliance on meat, is no exception.

In terms of the high fat content of ketogenic diets, fat is a concentrated source of energy with nine calories per gram. Prehistoric people were forced to eat when food was abundant and few parts of animals were left behind. Fat provided the staying power for days at end until the next capture or catch. In modern times fat still is very satisfying for hours on end—sometimes up to nine hours. Satiety from fat may prevent people from scrounging around for sugar- and calorie-filled foods and beverages.

Calorie counting (and often portion control) is not required in some ketogenic diets. This aspect may be very appealing to people who do not have the time nor the interest in doing it.

Perhaps one of the most important features of ketogenic diets is the lack of hunger. This is because ketone bodies tend to lower the level of *ghrelin* in the body. Ghrelin is a hormone that decreases hunger and promotes satiety. This is especially important between meals to help to manage appetite control.

The weight loss on a ketogenic diet may be more rapid than in a reduced-calorie or reduced-fat diet. This phenomenon may be very appealing to dieters because initial weight loss may be very motivational to continue dieting. When carbohydrates are limited as they are on a ketogenic diet, the body turns to its carbohydrates stores in the muscles and to liver glycogen. The initial weight loss on a ketogenic diet tends to be mostly water weight because when stored carbohydrates are broken down for energy, water is also released (glycogen is stored with water). After about two to three weeks of dieting, this water loss diminishes and fat loss increases, particularly when the diet is accompanied by exercise.

Protein, too, can be used for energy, but this is not the preferred fuel. The muscles, including the heart, are made of protein and these stores must be preserved. This is another benefit of ketogenic diets: most provide more than enough protein to help to "spare" the body's protein stores from use.

Another advantage of a ketogenic diet is its capability of improving serum triglyceride and high-density lipoproteins and potentially lowering the risk of cardiovascular disease. Diets of this nature are also capable of lowering the tendencies for elevated blood sugar, *C-reactive protein* (which is a sign of inflammation and a marker for certain degenerative diseases), insulin, and waist circumference.

Problems with Ketogenic Diets

Eating the same high fat and high protein foods and beverages day in and day out over time may become boring (like so many other diets), which may cause backsliding over time. Other criticisms of the ketogenic diet are that it narrows food choices, compromises nutrient intake, makes it difficult to eat out or socialize since alcohol and many desserts are prohibited, and may even lead to digestive issues.

Since the ketogenic diet is limited or void of carbohydrates with fibers (even fruits, vegetables, and whole grains), normal digestion and elimination might be compromised and a natural fiber supplement may be warranted.

Nutrient deficiencies may also be a concern on a ketogenic diet. Nutrients such as vitamins A, C, D, and thiamine, and minerals such as calcium, magnesium, and selenium may be lacking and might require supplementation. A health professional should be consulted before trying any diet—particularly one long term. Children, pregnant and nursing mothers, and the elderly should take special care and caution.

Coconuts and Ketogenic Diets

Research on ketogenic diets from the 1960s demonstrated that more ketones are produced by medium-chain triglycerides (MCTs) per unit of energy than other sources of energy. This is because medium-chain triglycerides are quickly transported to the liver via the hepatic portal vein as opposed to through the lymphatic system.

It follows that if a diet that is rich in medium-chain triglycerides but reduced in carbohydrates, then it might be effective in treating epilepsy and in weight control management. In fact, a diet with 60 percent medium-chain triglycerides with protein and some carbohydrates has been shown to provide more meal options for children with epilepsy. Dieters may also be able to tolerate fewer carbohydrates without boredom if a little variety is provided.

Coconut to the rescue! The range and depth of recipes and meals with coconut in all of its forms are mind-boggling, but for ketogenic diets, coconut oil with its high medium-chain triglycerides are key. Some recommendations suggest taking one to three tablespoons of coconut oil or MCT oil that is made from coconut oil daily. But this practice should support a ketonic diet, not replace it.

The Aging Brain

The brain ages like all bodily systems, organs, muscles, nerves, and cells. Specific changes in the brain due to aging include changes in blood vessels, free radicals, inflammation, neurons and neurotransmitters, plaques, and tangles and shrinkage.

Blood vessels in the brain may be compromised because arteries narrow with aging and there is less growth of new capillaries that occur. This may result in less blood being able to circulate and nourish the brain.

Free radical damage from cigarette smoke, herbicides, pollution, radiation, and a host of other environmental factors may accumulate with age. Free radicals are normally made in the mitochondria inside cells. They help the body's immune system to fight bacteria and viruses. However, free radicals may also damage the cell membranes of neurons or their DNA. This may instigate a chain reaction that releases more free radicals and cause further neurological damage.

Cells and compounds that are known to be involved in inflammation have been located in Alzheimer's disease plaques, so it is thought that inflammation may have a role in the disease. Inflammation increases when the body responds to abnormal situations, disease, injuries, or even stress.

Other changes in the brain are the result of changes in the neurons and neurotransmitters that are responsible for communications among neurons. If and when the white matter of the brain is reduced, then communication among neurons may be compromised. Structures referred to as *amyloid plaques* and *neurofibrillary tangles* develop both inside and outside of the neurons. *Amyloid plaques* contain largely insoluble deposits of seemingly toxic protein fragments. It is not conclusive whether these plaques cause Alzheimer's disease or whether they are by-products of the disease.

Neurofibrillary tangles are abnormal collections of twisted protein threads within the nerve cells that contain phosphate molecules. These threads may become so enmeshed that they may become tangled within the cells, disrupt the healthy neuron transport network, and damage neuron communication.

Decreased brain glucose metabolism has been identified before the onset of clinically measurable cognitive decline and may contribute to further cascading decline. The process of aging appears to increase the risks of deteriorating glucose utilization in some regions of the brain.

Alzheimer's Disease

Alzheimer's disease (AD) is a progressive neurodegenerative disorder that primarily affects the elderly, but can impact earlier in life. There are two types of Alzheimer's disease: *early* and *late-onset*. Early onset Alzheimer's disease may strike a person as early as their 30s, while late onset Alzheimer's disease may strike a person in their 70s or 80s and is the leading cause of dementia in people over 65 years of age. At the onset, people might experience memory loss and disorientation. As the disease progresses, additional cognitive functions may become impaired.

While there are theories that try to describe the course of the events that lead to Alzheimer's disease, the complete picture remains unknown. There are no totally effective prevention methods or treatments for Alzheimer's disease; however, there are some drugs that act as acetylcholinesterase inhibitors. *Acetylcholinesterase* is an enzyme that serves to terminate synaptic nerve transmission. These drugs enhance the effectiveness of nerve cells that are still functional but do not address the underlying pathology of Alzheimer's disease.

Dementia

Dementia covers a wide range of symptoms that are associated with a decline in memory and/or other skills that are involved in everyday activities. Alzheimer's disease is one form of dementia; it is estimated to account for about 60 to 80 percent of the instances of dementia. Another prominent type of dementia, *vascular dementia*, may occur after a stroke. Other conditions that may contribute to dementia are thyroid problems and vitamin deficiencies.

Signs of dementia may include the decreased ability to focus and pay attention; changes in memory, judgment, and reasoning; communication and language issues; and visual perception alterations. However, other environmental and physical factors may contribute to these signs of dementia.

Memory

Many cells in the body are formed and decline over a person's lifetime. Neurons develop throughout a person's life, but the brain reaches its maximum potential of neurons during a person's early 20s. They then slowly begin to decline in volume. Blood flow to the brain also decreases over time. However, the brain is capable of regrowth and learning and retaining new facts and skills. The more active that a person is, the more frequently their brain is stimulated. The better nourished that a person is, the better the odds that their brain will function well throughout their senior years.

Both episodic and long-term memory decline over time, as well as information processing and learning new information, doing more than one task at a time, and shifting focus to new tasks.

Coconuts in Ketogenic Diets and Brain Health

Coconuts with their medium-chain triglycerides have been associated with improved cognition and a number of enhanced brain functions. Studies have not definitively shown that medium-chain triglycerides can prevent dementia, but people with dementia might discover some short-term benefits by incorporating the medium-chain triglycerides like those that are found in coconuts into their diets.

Small clinical trials in aging individuals that had age-related cognitive decline and diabetic patients have shown that an MCT supplement can preserve cognitive function or lead to cognitive improvement. Medium-chain triglyceride supplementation appeared to help the diabetic patients but not the aging group.

Insulin, the hormone that regulates blood sugar, has been linked to changes in the brain that have been associated with Alzheimer's disease. This is the reason why this study was conducted with diabetic patients. It is not exactly clear what role insulin has in Alzheimer's disease. Furthermore, changes in the brain that are associated with dementia may be unconnected to glucose metabolism. Medium-chain triglyceride supplements also have been shown to improve cognitive function in people with mild cognitive impairment and Alzheimer's disease. A few studies suggest some cognition improvement and prevention of amyloid plaque formation in animals, but human studies have not been confirmed.

A recent study suggested that although the effects of coconut oil may be temporary, that both Alzheimer's and dementia patients had some short-term benefits from supplementation. Another study that examined the effects of a mild ketogenic diet during exercise, both with and without ketosis, demonstrated that endurance and cognitive function were both increased.

There is strong evidence that exists that the use of medium-chain triglycerides by healthy adults is considered to be low risk. Foods and beverages, such as coconut oil, that are high in medium-chain triglycerides have been widely used with few reported adverse reactions.

The medium-chain triglycerides with lauric acid in organic, cold-pressed, non-hydrogenated virgin coconut oil, which make up nearly 60 percent of the total fat content of coconut oil are preferable. For comparison: one tablespoon of coconut oil contains about 7.4 grams of MCTs; butter contains about 1 gram of MCTs; and palm kernel oil contains about 7.9 grams of MCTs.

In some clinical studies, 10 to 40 grams (0.4 to 1.4 ounces) of medium-chain triglycerides are ingested daily, although some reports indicate that a daily dose up to about 70 grams of medium-chain triglycerides may produce demonstrable results. Mild gastrointestinal side effects may be common in some individuals who consume medium-chain triglycerides. These side effects may be offset by consuming medium-chain triglycerides with food and by slowly incorporating MCTs into one's diet. Like other supplements or medications, it is important to discuss the usage of MCTs with a health care provider before executing.

Primary Considerations

A ketogenic diet must be followed very strictly because the body prefers glucose for metabolism. The simple addition of coconut oil to the diet may not necessarily provide the neurons in the brain with an alternative source for energy. The most optimal use of coconut oil for brain health may be its incorporation into a finely devised ketogenic diet. In this context, the rich concentration of medium-chain triglycerides in coconut oil will probably comprise a prescribed amount of the total daily fat intake and should be in the correct proportion to the amount of carbohydrates and proteins. An experienced dietitian/nutritionist or health care provider may be able to devise and monitor such a diet.

Coconut the Beautiful

Not only can people consume the goodness of the coconut, people can wear coconuts, too—on their face and body, that is! Coconuts have been extolled for their radiant properties in beauty and personal care. Coconut oil may be one of the best skin creams, hydrating serums, and moisturizing lotions available. That said, some of the magic-like qualities of coconuts are valid while others maybe more folklore. Let's explore!

The unique properties of coconuts make them extremely versatile and useful for a myriad of beauty and hygienic habits. This section will sort out the rights and wrongs for using coconuts for beautification purposes.

Skin Care

The cell membranes in the skin are made up of three fat-containing substances: glycolipids (lipids with a carbohydrate component that helps to identify cells), phospholipids (lipids with a phosphate component that provide structure and function to cells), and cholesterol (the waxy, fat-like substance that is found in all cells of the body).

Phospholipids, made of saturated and unsaturated fatty acids, are the largest component of cell membranes. The balance of these three fatty acids is important for proper cell function and it is critical to human and animal health. The fatty acids in coconuts contribute to a healthy balance of these fatty acids in the cell membranes of the skin.

Aging, heat and cold, medications, skin treatments, and the sun can affect the hydration of the skin in a multitude of ways. Replacing this hydration can be done both externally and internally.

Proper hydration is essential for the skin's cell membranes. The surface skin cells do not dry out as fast or as much and the dead skin cells are reduced or are easily eliminated. As a result, the skin's pores do not get as clogged with dirt and impurities and excess oils do not accumulate, which also reduces the possibility of acne and infections. Firmness and natural elasticity are generally retained and stretch marks and wrinkling also may be reduced in appearance.

Coconut oil provides hydration and essential fatty acids for healthy skin along with proper skin care. Coconut oil helps to protect the skin from free radicals and their aging effects and may help to improve the skin's appearance with its anti-aging benefits. Coconut oil acts as an antioxidant since it is stable, resists oxidation, and contributes some vitamin E, an antioxidant vitamin.

When coconut oil is absorbed into the skin and connective tissues, it may help to reduce the appearance of fine lines and wrinkles. It accomplishes this by supporting the strength and suppleness of connective tissues. It also aids in the exfoliation of the outer layer of dead skin cells, which makes the skin look smoother. Plus, it has a delightful aroma that bathes the skin to the enjoyment of the wearer and those in the immediate vicinity, so no other cologne or perfume may be necessary.

Skin-enhancing Qualities of Coconut Oil

Several of the skin-enhancing qualities of coconut oil and coconut products include:

Acne relief: Coconut oil helps to keep skin hydrated and the pores open to reduce impurities and acne-related skin conditions.

Age spot minimizer: Coconut oil may diminish the look of age spots when used on a regular basis. It may accomplish this by keeping the area of the age spots and also the skin around the age spots supple.

Anti-aging defense: By smoothing the look of wrinkled skin, coconut oil may help to make skin appear younger. Younger skin does have more moisture that may dissipate over the years.

Baby lotion/massage oil: Warmed coconut oil with essential oil such as lavender, mint, rosemary, or vanilla help to provide a relaxing, rejuvenating massage.

Bath/body moisturizer: Scented (or unscented) coconut massage oil can be applied to the skin both in and out of the bath or shower.

Body balm: Coconut-oil-based body balm is dense, hydrates, and repairs. It is similar to a lotion with light consistency that moistures but does not repair and a cream that is usually thicker, but is formulated for the face.

Body scrub: Coconut oil and coarse salt or sugar acts as an exfoliator for dry skin. Either fragrant (such a eucalyptus or lavender) or non-fragrant essential oils may be added.

Bruise reducer: Coconut oil may not directly reduce the appearance of bruises, but similar to its effects on scars, sores, and stretch marks, it may promote the healing of the surrounding skin over time.

Cold sore reliever: Cold sores, often due to viruses, may benefit from coconut oil's antiviral properties due to its composition of lauric and stearic fatty acids, know for their curative properties.

Deodorant: Coconut oil mixed with arrowroot powder, baking soda, cornstarch, and scented or unscented essential oils can be used as a natural deodorant.

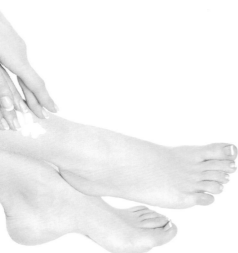

Diaper rash relief: Unless there is a known sensitivity to coconuts, a thin layer of coconut oil may provide some relief for chafed skin.

Dry feet remedy: Apply coconut oil to cracked skin and dry spots to smooth dry feet, especially in winter or after hot, dry summer sun.

Emollient: An emollient is a cream or an ointment that acts like a moisturizer. It is designed to help make the *epidermis* (outer layer of the skin) softer and suppler. Versatile coconut oil can be used as a base for homemade bath or body emollients or moisturizers.

Facial mask: The antioxidants and fatty acids in coconut oil help to lubricate, smooth, and soften the skin when it is mixed with baking soda, lemon, turmeric, or yogurt as a facial mask.

Lip balm: Coconut oil is a remedy for chapped lips since it is semisolid at room temperature and can easily be spread by fingertips. Coconut balm is particularly soothing for the lips. A touch of natural extract, such as vanilla, gives coconut lip balm a tropical flavor.

Makeup remover: Apply coconut oil directly to the face to cleanse it, or let it remain on the skin to rehydrate. Be particularly careful around the eyes so as not to over-oil them.

Night cream: Apply coconut oil at night for its maximum moisturizing benefits, particularly around fine lines or wrinkles. A little left on the pillow at night will also moisturize the hair.

Coconut the Beautiful

Nursing mothers' aid/nipple soother: Use coconut oil delicately around sore or cracked nipples to rehydrate, soften, and heal. This practice can be checked with an obstetrician or nurse midwife first before trying.

Shave cream: Coconut oil helps to provide a smoother, closer shave that leaves the arms, face, and legs evenly soft and moisturized.

Skin protection (dishpan hands, eczema, psoriasis): Coconut oil and coconut balm offer relief from allergic and chronic skin conditions that can be heightened by over-exposure to chemicals, sun, water, and other skin-damaging factors. Check with an allergist first before applying to very sensitive skin.

Stretch mark/scar reducer: Pregnant women and people with minor scrapes and scratches can use coconut oil as a topical treatment for marks and scars. Coconut oil probably will not directly cause fading, but it may help to prevent dark spots and/or blisters from forming. Coconut oil may also help the surrounding skin retain its moisture and healthy glow.

Sun block/soother: Coconut oil is soothing and has a natural sunscreen of SPF 4. It is still sensible to stay out of direct exposure of the sun. Sunscreens with a higher SPF should offer more protection from the sun's harmful ultraviolet (UV) radiation.

Under eye cream: Applied under the eyes and around fine lines, coconut oil combats a dried-out appearance and it is light and delicate for most skin types.

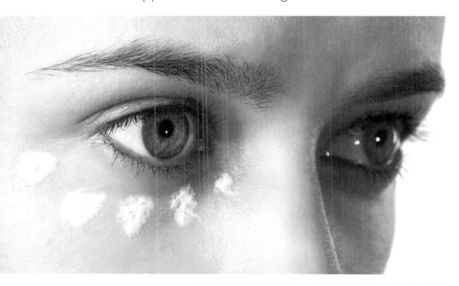

Hair and Nails

Like healthy skin, healthy hair and nails are formed, in great part, by a healthy diet that includes proper hydration. An unhealthy diet can lead to hair that is brittle, dull, or lifeless; a scalp that is flaky; nails that break and chip easily; and ragged cuticles. Coconut oil plays a role in radiant hair, smooth nails, and neat cuticles due to its fat content and along with coconut milk and coconut water helps hydration.

Coconut oil can be applied directly to the scalp, hair, nails, and nail beds. While excess coconut oil can be removed, the residue can remain for a healthy glow and moisturizing. Or, coconut oil can be mixed with some warm water, gently applied to the scalp, then allowed to soak into the scalp and hair before washing with shampoo and conditioner—coconut-based, of course! Plus, when coconuts are consumed they help to nourish the hair and nails from the inside out!

These are some of the uses of coconut products to enrich the hair and nails:

Adds shine and conditions hair: Coconut oil hydrates hair over time since it replaces the natural oils that shampoos tend to strip away. As a leave-in conditioner, coconut oil may condition the hair immediately and then benefit the hair over time. Some people think that application to wet hair is more desirable for sealing in moisture than on dry hair, but water and oil do not mix (consider vinegar and oil salad dressing), so it is best to experiment.

Dandruff control: Dandruff along with dry scalp may be chronic problems. The daily use of coconut oil literally confronts the root of the problem. A light head massage with warm coconut oil may also be invigorating.

Defrizzes hair and reduces split ends: Frizzy hair is often the result of dry conditions: too many chemicals in shampoos and conditioners, coloring, perms, or general over-processing. Coconut oil helps to calm frizzy hair and restore its luster. While regular trims are best for split ends, coconut oil may tame their unruliness in the short-term before the haircut.

Enhances eyebrows and eyelashes: Coconut oil may penetrate the hair follicles in the eyebrows and eyelashes and prevent a scaly look so that they appear to be soft and shiny.

Provides protection from ultraviolet (UV) radiation: Sunlight helps to create a pre-cursor of vitamin D on the skin's surface that is necessary for healthy bones and teeth, the heart, and immune function. But too much UV radiation may damage the skin and lead to some cancers. Coconut oil and its richness provide some measures of protection. But the wisest advice is to stay out of dangerous noon-day or tropical sun.

Reduces fungal infections: Coconut oil with its antimicrobial properties may help to prevent fungal infections of the nails (particularly toenails) and scalp. Coconut oil is even said to reduce candida, a systemic fungal infection. Make sure to check with a health care practitioner for specific advice.

Softens cuticles: Coconut balm or coconut oil soothes rough and ragged cuticles before they get out of hand. When either of these nurturing coconut products is applied to the nail base, they may enhance the nails and hands.

Stimulates nail growth: Nails need a moisturized environment in which to thrive. Coconut oil helps to prevent the nails from becoming overly brittle and split that may hinder their growth. Ample protein, biotin, B-complex vitamins, calcium, iron, magnesium, omega-3 fatty acids, and zinc are some of the nutrients that are needed for healthy nails.

Supports natural chemical balances in hair and nails: The body is a maze of natural chemicals and reactions that maintain its many functions. On low-fat diets, sometimes there is not enough fat or even protein for normal operations. Coconuts and their products may help support the body's quest for equilibrium of nutrient intake to keep it working smoothly.

Treats hair and scalp: Warmed coconut oil can be applied to the hair and scalp to treat dry hair before using shampoo and conditioner. The longer that it is left on the hair and scalp, the greater its moisturizing benefits. Stronger hair may be spared from dry ends that break or split up the hair and into the hair shaft.

Wards off infection: If used regularly, coconut oil may form a protective barrier on the skin around the nails to impede some common infections. Dry, cracked skin is more vulnerable. Make sure manicures are both safe and sanitary.

Muscles

Proper muscle tone requires a muscle-building and maintenance program. One needs muscle-building exercises, a regular exercise routine, and healthy muscle cells and surrounding tissues that support all of this muscular work. Of course, a good diet with protein and muscle-building nutrients such as B-vitamins, calcium, magnesium, and vitamins C, D, and E are vital.

The right kinds of fatty acids in sufficient quantity are essential to build and maintain healthy cellular membranes within muscle cells. Coconuts and coconut oil with their saturated fats supply some of this raw material. The size of muscles depends on genetic predisposition and on a dedicated muscle-building program, but a balanced diet supports these factors.

Cellulite, or subcutaneous fat within fibrous connective tissue, makes the skin look dimpled and nodular. Cellulite is common in the pelvic region, including the abdomen, buttocks, and lower limbs. Its causes may be genetic, hormonal, or lifestyle-related.

Effective treatments for cellulite are varied and debated. The use of coconut oil to reduce the look and texture of cellulite is one of these deliberations. The theory is that the human body requires certain substances to help to metabolize or break down its fat stores into energy.

The medium-chain triglycerides and lauric acid in coconut oil help to reduce inflammation, boost metabolism, and break down fatty accumulations. They work in conjunction with specific enzymes that break down fats into fatty acids and glycerol that are further broken down in the liver or used for energy. The enzymes are activated by the hormones epinephrine, glucagon, and growth hormone.

In the process of using these medium-chain triglycerides and lauric acid, the skin may become more elastic, silkier, and smoother in appearance and the bumpy, holey look of cellulite may be diminished.

Spider veins are small and twisted blood vessels that may be visible through the skin. They may look more pronounced when the skin is dry or saggy in appearance. Coconut oil can be gently massaged into the areas where spider veins appear on the face or legs. They may help to restore the surrounding cells and detract from the redness.

Teeth and Gums

The fat-soluble vitamins A, D, E, and K require fat for proper absorption and assimilation. In particular, vitamins A and D are essential for strong bones and teeth. By remineralizing tooth enamel, these vitamins help to build up the protective qualities of the teeth and help to guard against cavities. The anti-inflammatory, analgesic, and antipyretic (fever-reducing) properties of coconut oil may help to defend the teeth against decay. Healthy teeth and gums are one of the body's first lines against infection.

In particular, coconut oil:

- **Helps prevent periodontal disease and tooth decay:** Due to coconut oil's anti-inflammatory abilities, it may help to keep the mouth and gums healthy and even fight infection.

- **Improves calcium and magnesium absorption and supports the development of strong bones and teeth:** Coconut oil with its medium-chain triglycerides may have the capacity to increase calcium and magnesium absorption, which in turn may help to promote strong bones and teeth. Many other dietary factors are still in play, such as calcium, phosphorus, potassium, and vitamin A, C, D, and K intake.

- **Whitens teeth and acts as a mouthwash:** With its antifungal and antibacterial properties, coconut oil that is mixed with a little baking soda makes a natural whitening and protective paste. A few drops of an essential oil, such as anise with its licorice taste or mint or may be added.

Sun Damage and Aging

Sun damage is due to sun exposure throughout the years from everyday activities to conscious sun bathing. The changes in the earth's ozone layer are also responsible for undesirable ultraviolet (UV) radiation. Fair-skinned and darker-skinned individuals are both prone to sun damage; however, fairer-skinned individuals are more at risk.

In some parts of the world where there is intense and nonstop sun exposure, residents do not seem to have the same type of sun damage than they do in other locations. It may be that their local diets are more abundant in antioxidant-rich foods and vegetables that feed and protect the skin from the inside out.

Cellular degeneration may be caused by *free radicals* in the environment. Free radicals may be formed as the result of exposure to sunlight and air pollutants, but they may also be due to alcohol consumption, diet, drugs, exercise, inflammation, smoking, and one's diet.

One method to delay or prevent cellular degeneration from free radicals is with a diet that is rich in antioxidants (substances that help to halt oxidation which causes cellular destruction). Coconut oil has some antioxidants, but it is more valuable for its saturated fatty acids.

A diet that is high in polyunsaturated fatty acids may lead to cellular membrane instability and make one more prone to cellular damage by the sun and environmental factors. Saturated fats are more protective.